THE ORIGIN

OF THE UNIVERSE

THE ORIGIN
OF THE UNIVERSE

JAN ALEKSANDER PIASECKI

PHILOSOPHICAL LIBRARY
New York

71337

Published, 1972, by Philosophical Library, Inc.,
15 East 40th Street, New York, N. Y. 10016
Copyright © by Aniela Piasecki 1971
All rights reserved
Library of Congress Catalog Card Number: 72-171468
SBN 8022-2071-1

The Origin of the Universe was written in 1954 by
Jan Aleksander Piasecki. Translated from the Polish
original by Terenia A. de Rzonca, 1969-70.

Manufactured in the United States of America

A NOTE ABOUT THE AUTHOR

The time in which Jan A. Piasecki arrived into this world was an epoch without hope for Poland. He was born in the town of Mielec in the western portion of Poland. *When he was barely five years old he lost his parents.* He was brought up by his grandparents. He attended school in the town of Mielec. After school hours he wandered along the banks of his beloved river "Wisloka". It was probably in this beautiful and lonely place that he communicated with Nature and started to acquire his deep insight. It was here he contemplated Nature and pondered over its creation and the Power behind it.

At a later time and distant place his thoughts were developed into his theory, "THE ORIGIN OF THE UNIVERSE," although he avoided the technical language as much as possible, so that the general public could easily understand it. He read many books and he was mostly interested in the history of his Country, and shared the yearning of his countrymen for freedom, after a century and a half of partition. He also dreamed of the reconstruction of a Poland: pure, just and democratic as she was prior to the acceptance of Christianity.

When he was 16 years old, the war of 1914 erupted. His first thought was—freedom for his Country!—Without any hesitation he joined the Polish Legions under Marshal Józef Pilsudski, and he was part of the famous "FIRST BRIGADE" of which to this very day songs are sung in the Land of his fathers.

He was brave but was never offensive. His love of honor was greater than his fear of danger. He did not care about material wealth and was rather altruistic. He respected all people and was in turn respected by them.

In the army he distinguished himself repeatedly, showing great self-discipline. Guided by youthful ardor for the cause of freedom, he persevered with the small army of young defenders, merely children who lacked food and arms, and who were forced to fight for the life of their Country against the huge, well trained and well equipped army of the enemy. And their great spirit saved the Country.

However, the new found freedom for resurrected Poland did not last long because the "Blitz-Krieg" of September 1, 1939 started a new world conflagration. The Poles again had to shed blood and die for freedom. But this time Jan A. Piasecki, to his regret, was unable to defend his Country.

He left Poland for the U.S.A., in 1921 after having been in Italy, France and other European countries. He wished to see the North American continent: its people and how they live.

The same strength of character and self-discipline that served him in the past directed him to acquire a trade and master the English language. He studied design, and was graduated from designing school and worked as a dress-designer until 1956. At this time he fell ill and moved to the country to regain his health. In the fall of 1966 he died.

His last ten years were productive, devoted to contemplating and writing poetry and gardening.

He had considered all of his work not as a means of making a living but as an art. He also was a great chess player, the only game he respected.

Above all, he had completed this book which I am trying to publish in his name as a memorial to him.

Aniela Piasecki

(his wife)

CONTENTS

THE ORIGIN

OF THE UNIVERSE

INTRODUCTORY NOTE

The universe has always been a mystery to us. It is still an enigma today, despite scientific and technological achievements. As soon as Man became aware of his environment, fear of it thrust him to seek an explanation of it. Throughout the centuries, Man has attempted to make sense of his world and to devise methods of easing the burden of his humanity. The author has made sense of the universe in this thesis. He has provided us with a window looking onto the universe through which we can observe the marvelous workings of Creation.

It may be helpful to the reader before discussing this thesis, to briefly examine the methods of science. This should then place the theory proposed here in a clearer perspective. The scientific method requires observation of Nature and experimentation with it to determine the relationships between the items under test and to relate them mathematically. To illustrate, consider the relationship between pressure and temperature of a gas in a sealed vessel. The higher the temperature, the higher will be the gas pressure of the constant volume of gas. When this relationship is noted, the scientist will test this phenomena over a range of temperatures and the mathematician will incorporate the test results into a mathematical formula. This then becomes an experimental law of Nature and, if the exact conditions of the test are specified, this relationship will hold throughout the universe. Other relationships of this type quickly come to mind, such as the flow of electric currents, reflection and refraction of light, etc. If science were only limited to discovering laws of Nature, we would not be much further ahead today in our knowledge.

For science must also answer the question why. Why does Nature behave in this way only and not in any other way? Scientific theories generally provide this answer. A theory usually but not necessarily combines several or many of the laws of Nature, offers an explanation of why Nature behaves as it does in these particular cases, and generally throws in some complex mathematics to go with the explanation. Theories are the ultimate creations of the human mind.

How successful have we been in our quest for understanding through science? Have our scientific theories stood the test of time? Each generation has produced their own theories consistent with experimental evidence of their time. As new investigation is attempted and new evidence accumulated, existing theories are expanded, modified and then abandoned. A theory is discarded because it could not explain new experimental facts or observations. Science is then left in a state of confusion until a new theory is proposed. This is what happened when Newton's theory could not explain the new experimental evidence at the turn of the century and was superseded by Einstein's theory of relativity. Scientific theories have changed and will change as further scientific knowledge is accumulated. Although progress has been made in science, a word of caution is in order. The present state of science does not give us a complete explanation of the universe.

The writer of this essay, "The Origin Of The Universe," has approached the enigma of the universe through observation, meditation and logical analysis of the data which he observed in the world around him. His was a philosophical approach, rather than an approach encumbered with mathematical and experimental data. His approach was original, drawn from within himself and from material phenomena. Further, he believed in the consistency of Nature, i.e. what was true of his observations must apply to all Nature and to the whole universe. Yet, can we be aware of any phenomena which cannot be sensed, which the least of us, even a child, can sense? The answer to this is, no. We may see a great deal more by using telescopic and/or other instruments but, if the

4

universe is consistent, then the same laws apply to what we can see and touch and to the world as seen through these instruments.

This book is written in simple terms, intelligible to the average person. Mathematics is not used but simple, direct descriptions of universal phenomena are given. The writer has avoided the use of scientific terms and in the introduction, defines the terms used clearly. It is unfortunate that in this day, scientific theories are written in such abstract terms with mathematics generously interspersed, that only another scientist of the same discipline can understand it. This is not the case here.

The theory proposed here has been developed from a synthesis of data accumulated over the years by the author. The theory is short and highly concentrated, leaving to others its scientific expansion and philosophical implications. The theory discusses the forces in the universe and how they combine and operate to create the universe and matter, and all living things in it. There is a special section on the soul, its creation and its destiny. The theory ranges over a universe of objects and it is all-inclusive in its scope.

What is the origin of the moon? This is the question under scientific investigation by the Apollo Program of flights to the moon. Existing theories are that the moon was once part of the earth and was separated by some cataclysmic event, or that it was created at the same time as the earth from the same pool of matter, or that it was a planet trapped by the gravity of the earth. These pages will also provide the answer to the origin of the moon.

A. P.

5

INTRODUCTION

Entities are everywhere, in light and in shadow;
Always present, and in every creature.

Whatever is created, is created in the midst of never ending change. Whatever and wherever it was, is and will be created; it was not, is not and will not be created through a miracle, nor chance, nor without a goal. All forms on each planet and throughout the universe are created from ENTITIES, by ENTITIES, and governed by ENTITIES during the entire time of their existence. We feel with the feeling of ENTITIES, and we differentiate ENTITIES by the feeling of each respective ENTITY.

In order not to keep the reader who is interested in my theory in the shadow of hypothesis, I shall immediately clarify the nature of these ENTITIES. They are two temperatures, that is: HEAT and COLD. These are the two: POWERS or ENTITIES.

Therefore, when I think of space, I come to the conclusion that it is something which not only cannot be created, but cannot be thought of as if it never did exist or could some day cease to exist. Through my imagination I can remove the universe from space and leave in it absolutely nothing; but I cannot remove space with my imagination. For my perception of it tells me of neverending infinity in time and space.

In the same manner, I can only think of two ESSENCES which fill in completely so called "space" and all forms of the universe, whether already created or in the process of being created. Also, I cannot visualize in any way whatso-

7

ever, that even the smallest or the minutest of spaces be not filled in, if not with Both, then with one of Them. So that, what we call "space" is not an empty expanse in between forms of the universe, as it may seem from mere appearance. Now then, the two ESSENCES which completely permeate infinite space and whatever is found in it are ENTITIES. By ENTITY, one should understand that it is something which is not made up from any particles, that is, atoms; nor does it contain any mixtures.

Consequently, one ENTITY alone does not fill in infinite space, nor does it create from itself. Because, if one ENTITY only filled infinite space, there would not be any universe in the latter. But both ENTITES create together within each other unlimited forces and power to create, through never-ending changes in their state or condition of pressure and hardening.

Each new changing state in one ENTITY as in the other— of which states there is an infinite number, from the mildest to that of unlimited power—has its task and governing power to create all forms on each planet and in the entire universe.

The ENTITIES when compared at similar stages are equal in infinity, size and power. Each ENTITY retains its individuality, wherever and in whatever it happens to be present. And each one knows itself and knows the other like itself, and the opposite characteristics of the other. Both Powers are knowledgeable of their mutual powers of creating, governing, and guiding unlimited number of different forms on the planets and in the universe.

Consequently, both ENTITIES—in spite of their infinite expanse—are always in constant movement resulting in hardening and pressure; and never and nowhere do they follow the alternate negative way of dissolving or dilating into each other. Never does their state remain the same, even during the blinking of an eye. But through mutual pressure and forcing through oneself, and because of their equally balanced POWER, they never and nowhere confront each other.

The impact and expansion of the HEAT ENTITY and the resistance and pressure of the COLD ENTITY are the main causes of all movements in the ENTITIES and in all that they create.

The HEAT ENTITY draws the strength for its impact and expansion from its inner tension. Its speed and strength of impact depend on the degree of its tension. And the power of its tension depends upon the pressure applied by the COLD ENTITY.

The greater the resistance of the COLD ENTITY, the greater the tension is built up within the depths of the HEAT ENTITY.

The HEAT ENTITY immediately lets go of its inner tension the moment the power of its tension begins to overcome the surrounding powerful pressure of the COLD ENTITY.

The HEAT ENTITY could never be contracted into any substance.

The HEAT ENTITY creates within itself the nucleus for all particles, that is atoms.

The COLD ENTITY draws the strength of its resistance and pressure from the hardening within itself. According to the greater or lesser hardening within itself, the greater or lesser becomes its resistance and pressure. Its power to harden depends on the impact and tension of the HEAT ENTITY.

The COLD ENTITY hardens most on the layer touching the HEAT ENTITY. The more the COLD ENTITY hardens, the more it contracts. It can harden into an infinite number of hard substances. The COLD ENTITY creates within itself the outer covering of all particles, that is, atoms.

THE SUN

Just as the ENTITIES did not have a beginning nor will they ever have an end, whether in time or in their unlimited greatness, so the universe did not have in them a beginning nor will it have an end, only the *forms* in it constantly change.

For every single *form,* beginning with the largest Sun that ever was, is or will ever be created, to the minutest, primeval particle, that could ever be created, has its beginning, transformation and end. And from that which it was created, to that it finally returns.

If within the boundaries of constellation, the Sun, planets, moon and the remainder of the forms within it would dissolve — not into particles, but into the very ENTITIES — to what degree the hardening and pressure within the ENTITIES would then arise in that given area of space?

If in that given area of space the hardening and pressure within the ENTITIES would rise to the point of tension, because of the transformation and dissolution of the forms within the universe, then the ENTITIES would begin to create a Sun from the HEAT ENTITY.

With the increase in power from the hardening of the COLD ENTITY and the tension within the HEAT ENTITY, the HEAT ENTITY, because of the energy released through its loosening, forces itself through the hardening COLD ENTITY by means of huge waves. Some of the waves detach themselves from the main HEAT ENTITY and start forcing their way on their own into the COLD ENTITY. The stress of the waves all around the COLD ENTITY, and, especially, in front of them heighten the hardening of the COLD ENTITY.

The increased hardening within the COLD ENTITY forces

11

the waves to contract within themselves and to narrow down. The contraction and narrowing down of the waves increases their speed. Their speed intensifies the impact of their wave fronts into the COLD ENTITY. Whenever one of the waves encounters resistance at its front which causes it to reduce its speed, then the rest of the wave presses down, mostly at the front, with such pressure that its entire form becomes visible, — that of an elongated wave with a bright front. And this form of the HEAT ENTITY races through the COLD ENTITY until it collides with another similar form to itself, or until it becomes restrained to a stop by the COLD ENTITY.

From the moment the forms collide at their fronts, the powerful impact and the lightning-quick arrival of the remainder of their wavelength forces the center to rotate and wind itself from this remainder into a gigantic ball. As this remainder is being wound, the COLD ENTITY, pressing from all sides, hardens itself into an unusual density around the Sun. If one of such forms would speed ahead with such velocity which would cause in front of it such hardening in the COLD ENTITY, which would stop it at once, then it would also wind itself into the shape of the Sun.

The HEAT ENTITY compressed into the shape of the Sun — through energy released by its expansion — pierces through with thin, contracted, dense and uniform rays the encircling and much hardened layer of the COLD ENTITY.

The heat rays speed ahead in straight lines equi-distant to each other; their speed depends upon their diameter and the intensity of the pressure within the Sun's (HEAT ENTITY'S) core.

Such being the case, as the sphere advances through space ahead of speeding rays, the rays do not change their equi-distance to each other; but with a mild, gradual decrease in the COLD ENTITY's density in the direction of their path, and expanding from their inner tension, the rays increase their volume in a uniform manner. The gradual increase of their volume regulates in turn the hardening within the COLD ENTITY.

The hardening within the COLD ENTITY keeps the rays at an equal distance to themselves. The Heat rays speed ahead until they collide with some other form in the universe. At the very moment of their collision, they explode, disintegrate and unite into an Entity. The rays, who do not encounter other forms in their path, speed ahead until they reach and enter their proper HEAT ENTITY.

The pressure of the COLD ENTITY on the surface of the Sun restrains the rolling waves upon it. The restraint of the waves presses and contracts the HEAT ENTITY in the center of the Sun, diminishes its circumference and increases its rotation. The diminished circumference of the sphere increases the pressure of the surrounding hardened layer of the COLD ENTITY. The pressure of the COLD ENTITY and the increased rotation of the Sun, restrain the contracted HEAT ENTITY from a violent disintegration and from powerful, straight-hitting, massive blows of the surrounding and greatly hardened layer of the COLD ENTITY.

The horizontal blows of the surface Heat waves no longer have a powerful enough impact in order to penetrate the much hardened crust of the COLD ENTITY.

MOLECULES

While the hardened layer of the COLD ENTITY is being swept by the rolling and swirling waves on the surface of the Sun, the ensuing friction causes splinters—into which the HEAT ENTITY particles have inserted themselves — to break away from it and roll down into round or oval molecules. If several such particles that broke away from the HEAT ENTITY, would fuse together into one thin piece, then in the process of rolling into a molecule, the HEAT ENTITY particles would unite within it into one nucleus. However, the majority of the primordial molecules from the surface of the Sun are rubbed back into the compressing surface of the COLD ENTITY.

These rubbed in molecules along with the fused particles from the HEAT ENTITY are torn off along with the splinters and rolled into new molecules. The second molecules may contain within themselves one or more primordial molecules. In the ensuing creation of multiples, the process of fusion, tearing off, and rolling of molecules keeps repeating itself. Let us take for example, a triple molecule: this multiple molecule may contain single molecules of different elements or one multiple molecule made up of different elements.

In every multiple molecule there are primordial molecules present. The number of times the creating process of fusion and rolling repeats itself depends on the size of a given multiple molecule, and on how easily its outer covering can be torn off. Therefore, every kind of molecule has its own individual nucleus acquired from the compressed HEAT ENTITY regardless whether it is a complete multiple molecule or whether it is within, among others, another multiple molecule.

In spite of the powerful compression of the HEAT ENTITY

15

within the Sun, all newly-created molecules are difficult to split open. Taking into account the number that are created, very few crack, because their outer coverings are thicker and because they are propelled by an ENTITY and within the HEAT ENTITY.

THE TRANSFORMATION

OF THE SUN INTO A PLANET

When the surface of the Sun has condensed into a thick viscose substance from the various molecules — the molecules having been propelled upward by stormy surface spirals — the molecules collide with each other, coagulate and harden into solid masses of various shapes and sizes. These masses in turn attach themselves to each other and become grouped into one solid mass by means of the pressure exerted by the hardened layer of the COLD ENTITY around the Sun.

The steady escape of the HEAT ENTITY's rays from the Sun and its constant use into the making of the molecular nuclei; its compression within the Sun's core by the pressure of the COLD ENTITY and the formation of a thickening crust on the Sun's surface, causes its circumference — rather the circumference of the new planet — to be several or many score times smaller than the original sun.

Molecules are constantly being created without interruption during the entire existence and transformation of the Sun and during the entire span of the HEAT ENTITY's vortex spiralling within the planet's core.

During the entire existence and transformation of the Sun into a planet, the HEAT ENTITY, compressed within the Sun, does not stop, even for a moment, its turbulence and spiralling — the violence of its huge fountains tumbling and propelling upward various shaped masses. These masses, under the pressure of the COLD ENTITY, fall back on the planet's surface. But some masses, propelled with a mightier thrust, piercing through the surrounding viscose ring of the COLD ENTITY, wander in

17

space until they reach another sphere of the HEAT ENTITY. And when they enter it, they disintegrate into molecules, or fall upon another planet.

However, a planet can never be formed from such cast out masses. Every real planet, must have within itself a spiralling core (nucleus) of a compressed HEAT ENTITY. Therefore, such an outcast mass, even though it were the greatest in size, and even though it would have within itself a small portion of the compressed HEAT ENTITY, in the course of its journey, however, this small portion of the HEAT ENTITY, freeing itself from its compression would escape in between the molecular spaces of the surrounding COLD ENTITY before this mass could ever reach another sphere of the HEAT ENTITY; and, therefore, upon entering it, it would then disintegrate.

The entire surface of the Sun — in the process of its transformation into a planet — is covered with unleashed, flaming tongues of the HEAT ENTITY escaping from beneath the crust of the newly-forming planet. An uncounted number of brilliant streams zigzag through the upper strata which has become dense with loose molecules and groups of molecules. The planet remains in such a turbulent state until the HEAT ENTITY escapes beyond the surrounding strata and out of the entire planet's crust, with the exception of larger cubicles in which some portions of the HEAT ENTITY remain sealed, for a time.

As the HEAT ENTITY escapes from the planet's crust, the COLD ENTITY fills in with itself the now vacant spaces in between the molecules and any other that may occur within the planet's crust. Upon the escape of the HEAT ENTITY from the planet's crust and its upper strata, all molecules, whether in groups or loose, which happen to find themselves in the upper strata, begin to fall down — due to the pressure of the COLD ENTITY — to the surface of the now newly-formed planet.

On the surface of the planet there now are: solid masses of various shapes and sizes, gigantic cliffs and ravines, huge holes caused by escaping portions of the HEAT ENTITY which were cut off from the planet's main core and sealed within cubicles of the COLD ENTITY's crust while the HEAT ENTITY was escaping and compressed under the COLD ENTITY's pressure and the thickening of its crust.

And so, the compressed force of the HEAT ENTITY within the planet's core spirals within itself in order to expand, and so rotates the sphere of the newly-formed planet.

The new planet, now with a compact solid surface permeated with the COLD ENTITY — but without atmosphere — begins to seek a newly-formed sun. For during the time that a sun transforms itself into a planet, somewhere in space a new sun is being born.

TRANSFORMATION OF PLANETS INTO MOONS

During the entire span of a Sun's existence and its transformation into a planet, each planet in its solar system having an atmosphere becomes used up through the friction and splitting of molecules upon it, and the escape of the HEAT ENTITY from its core.

From the moment that the Sun's rays begin to stop because of the formation of a crust upon its surface, all living forms cease to exist on the planets that form part of the solar system of the Sun now undergoing transformation into a planet.

Towards the end, when the Sun is almost entirely covered with a crust, the remainder of the HEAT ENTITY left within each planet's core begins to erupt violently. The planet's rotation slows down; their crusts split open; the waters become turbulent, steam, and press through the fissures of the planet's splitting crust.

The HEAT ENTITY, piercing through the weaker walls of the planet's crust, breaks, disintegrates, and propels upward whatever is in its path unto the planet's surface.

All loose molecules from the planet's crust and in the upper strata, resulting from this disintegration, split open under the HEAT ENTITY's impact. The molecules not subjected to this breakage liquify and spread over into a compact layer which eventually hardens into a crust.

After this great eruption upon the planets and after the dissolution of the HEAT ENTITY into the upper strata from its powerful compression, the surrounding COLD ENTITY, permeating the planet's crust, greatly hardened on its exterior and interior because of the HEAT ENTITY's escape from the planet's core and from the nuclei of the splitting molecules, then, the

21

COLD ENTITY, relaxing from its hardness, expels the remnants of the escaping HEAT ENTITY from the planet's crust.

The COLD ENTITY — permeating the vacant spaces in the planet's crust, helped by the dissolution of the HEAT ENTITY's molecular groups which roll down together with the crust's broken-down pieces into crevices and wider fissures — compresses and pulls down the planet's crust.

The planet sphere, now transformed into Moon, permeated and hardened with the COLD ENTITY, stops rotating. There is no water, nor atmosphere. Its size is diminished several times from the original one. Its entire surface is compactly covered with a very hard mass consisting of thick and toughly coated molecules with smaller and less compressed nuclei. Molecules belonging to this category, linked with one another, no longer rotate, and only have the ability to adhere and compress themselves into hard substances. Such molecules exist so long until the COLD ENTITY does not dissolve away from their coating, or crust.

Therefore, the new moon from a transformed planet drifts towards a newly-formed planet from a transformed Sun, and becomes a COLD ENTITY reservoir so as to maintain a balance on the planet among the ENTITIES. The COLD ENTITY, surrounding whatever space is occupied by the moon, fills in all the inter-molecular spaces within it.

Before the Sun is completely covered with the molecular crust, the entire surface of its planets will be covered with water and their upper strata filled with a thick fog. If the crust of a planet transforming into a moon could not endure the violent blows of the erupting HEAT ENTITY from the planet's core and from the nuclei of the splitting molecules, and would as a consequence explode into pieces, then its pieces, namely, meteors, would wander individually until the COLD ENTITY would dissipate itself from the molecular crusts, or until some of them would fall upon some planet.

An old moon, which belonged to a planet now transformed into a moon, and which has not disintegrated under the impact of the Sun's rays, or in which the COLD ENTITY has not dissipated itself from its molecular crusts during the span of the Sun's existence, such a moon — or rather, that which is left of it — wanders alone, and its end is the same as that of meteors.

It may happen that not all planets in a Sun's Solar System would succeed in creating their atmosphere before their Sun's transformation into a planet. If such a planet did not succeed in creating its atmosphere before the Sun became covered with crust, such a planet would not transform itself into a moon, nor would its core escape from it; but, it would drift with its moon or moons along with the planet of the newly-transformed Sun towards a newly-formed Sun somewhere in space.

When the planet seeking a newly-formed Sun approaches it at the distance where the impact of the Sun's rays hitting it over-powers the force of the planet's speeding impetus, then the planet deviates from its course and begins to orbit. So that, from one side, a given planet is stopped and pushed backwards by the impact of the Sun's rays, and on the other side, it is pressed upon by the hardened COLD ENTITY. Therefore, such a planet is forced to revolve in a corresponding orbit and distance from the Sun.

(However, one must remember that a planet in space is entirely weightless.)

Hence we know that the Sun is the source of the HEAT ENTITY in the maintenance of constant movement among the planets of its Solar System.

CATEGORIES, SPECIES AND KINDS OF MOLECULES

The primordial molecule, created from both ENTITIES, is the first and smallest individual moving form on the planet, and the beginning of all living forms upon it, for they are all created from it.

All the planet's molecules are divided into two categories: the Heat Category and the Cold Category. A molecule belongs to the Heat Category if its nucleus which was formed from a torn-off particle of the HEAT ENTITY is greater in size than its crust which was formed from a torn-off particle of the COLD ENTITY. On the other hand, a molecule belongs to the Cold Category if its crust which was formed by a particle from the COLD ENTITY is greater in size than its nucleus, which was formed by a particle of the HEAT ENTITY. The latter molecules are not easy to break up. On the other hand, the Heat Category molecules are easier to break up.

Hence, the Heat Category molecules have the attributes of the HEAT ENTITY; and the Cold Category molecules have the attributes of the COLD ENTITY.

Each Category also contains two species of molecules: the primordial molecule and the multiple molecule. The primordial molecule belongs to the first species and the multiple molecule to the second species.

The second species occurs when the primordial molecules are rubbed back into the COLD ENTITY's crust and then torn off along with the splinter — into which were infused particles of the HEAT ENTITY — and then rolled into the first multiple molecules. Other kinds of multiple molecules are formed according to a previously mentioned process.

Every primordial molecule of the Heat Category — which

25

is the class of the minutest molecules that could ever be created, — has a COLD ENTITY crust and a nucleus from the smallest particle of the HEAT ENTITY. Whereas, the HEAT ENTITY in its nucleus predominates over the COLD ENTITY particle in its crust more than in any other molecules, consequently, such a molecule has the thinnest crust and a nucleus with the weakest compression; it splits easily; has a weak explosion; rotates more speedily; is torn off the quickest; is more volatile; and hardly has any tendency towards cohesion. Such a molecule is one of the most important components of the "atmosphere", that is, the air.

The task of such molecules is as follows: to loosen and break up molecular groups into single molecules; to regulate and maintain the proper and necessary elasticity within organic bodies in order to maintain them in constant movement through the splitting of the molecules within them, and the dissolution of the HEAT ENTITY compressed within the nuclei.

The following gradual evolution of the various classes pertaining to the Heat Category — including the last primordial molecule on this scale in which the HEAT ENTITY particles have the superiority over the COLD ENTITY particles — ends with the beginning of the next Category, that is, the Cold Category.

Each molecule in each progressing class of the HEAT CATEGORY is formed with a larger particle of the COLD ENTITY in its crust as well as of a larger particle of the HEAT ENTITY in its nucleus, until the last molecules on the scale of this Heat Category are formed. In these last molecules, the superiority of the HEAT ENTITY particle gradually diminishes, until in the very last molecule it is hardly perceptible. In the following molecules—which begin the Cold Category—the COLD ENTITY particle will gradually begin to have greater superiority over the HEAT ENTITY particle.

The primordial molecule at the end of the scale of the Heat Category and at the beginning of the Cold Category, contain bigger particles of both ENTITIES than any other primordial molecules and are the largest of all primordial molecules.

They are the main components of water and of the atmosphere, that is, the air.

Some of them may even have a greater compression within their nuclei than it is within the center of the Sun itself.

Therefore, the primordial Cold Category molecules will evolve in the same manner as the primordial Heat Category molecules, with the exception that the COLD ENTITY particle in its crust will outweigh and be superior to the HEAT ENTITY particle in its nucleus.

It follows then that both Categories have an equal gradual classification of primordial molecules: the HEAT ENTITY particle in the nucleus of the Heat Category primordial molecule is equal to the COLD ENTITY particle in the crust of the Cold Category primordial molecule. Also, the HEAT ENTITY particle in the nucleus of the Cold Category primordial molecule is equal to the COLD ENTITY particle in the crust of the Heat Category primordial molecule.

It follows further—assuming that Both ENTITIES were at equal states— that in all primordial molecules and in all ensuing molecules which form the planet's crust, the number of particles torn off the HEAT ENTITY is equal to the number of particles torn off the COLD ENTITY.

The primordial Cold Category molecules, evolving at the end of the scale of that gradual classification, have the greatest tendency toward cohesion; they are almost completely motionless; and they are used up very slowly—in comparison to others—during the entire span of the planet's existence.

All multiple molecules are subject to the same classification and characteristics as the primordial molecules. It doesn't matter where they are found; whether they are inside other multiple molecules, or whether they are themselves complete multiple molecules.

Therefore, to sum up, all molecules belonging to the Heat Category— through the force of their rotation which they draw from their nuclei, that is from the compressed particles of the HEAT ENTITY— restrain and prevent the Cold Category molecules from attaching and adhering to each other

27

into solid masses wherever and in whatever form they may occur. All molecules of the Cold Category—by the force of their attraction and cohesion which they draw from their crusts composed of the hardened COLD ENTITY particles — regulate the rotating velocity of the Heat Category molecules, and restrain their dissolution and escape from wherever they happen to be.

INTERMOLECULAR SPACES

If one should happen to look at a pile made up of hard, round and of equal size balls, one would notice on the outside of the pile openings to spaces which surround on all sides the balls making up the pile. If one would arrange balls of equal diameter in a pile so that each ball would be surrounded with twelve balls, one would notice that the area where the surface of the inner ball is touched by the other balls, that this area is smaller than the open space around the ball.

Therefore, each ball in a pile and everything which is made up and formed from molecules has within itself a similar surrounding space in between molecules, according to the size of molecules and the group of molecules. Therefore, the intermolecular space is greater or smaller according to the molecules.

However, no molecule—not even the smallest—can enter this intermolecular space in between a group of molecules without pushing them apart, if the molecules were of equal size and arranged at equal distances, and even though this group would be made up of the largest multiple molecules.

And so, just as throughout the infinite expanse there are no empty spaces, so on Earth and within the Earth there are no empty molecular spaces in anything or anywhere, even though the object would be made up of the minutest primordial molecules that could ever be created.

All intermolecular spaces, wherever and in whatever form they are found, are always compactly filled with the EN-TITIES that press through them. So that every single individual molecule on Earth—no matter where it may be

29

found—is always constantly surrounded, if not with Both, then with one of the ENTITIES. In this manner, the ENTITIES have dominion and power over every single molecule on Earth and wherever it may be found. And so, the molecules are Their first bricks of matter from which They constantly build new shapes and forms on the planets.

A PLANET IN THE SUN'S ORBIT

From the moment a newly-formed planet—without atmosphere and water—enters a Sun's orbit, its hard surface begins to crumble in the following manner. The beams' constant and forceful blows on the planet's surface and the ensuing heightening of the Heat Entity's pressure upon it up to a visible state cause the disintegration of these beams after their impact, which impact in turn loosens and crumbles the planet's hard surface.

The planet's surface, subjected to the straight blows of these beams, in a given time, could be compared to the surface of iron melting in a foundry. Its entire surface is steaming with a variety of loosened, simple, volatile molecules.

At the outset, before the planet's atmosphere became filled with molecules, the planet's entire surface was dark. When it was in the vicinity of the sun's rays, the planet appeared as bright as a full moon. But, as its atmosphere started to fill up with molecules, the planet became brighter because of the Sun's beams which constantly bombarded these molecules and pressured them into the Heat Entity until they became visible. As these molecules rotated uniformly in the atmosphere, they threw off sparks of light upon each other, thus creating a degree of brightness. This atmospheric brightness, at this stage, could be said to be reddish-grey.

One can surmise, that during this period when the planet's atmosphere was being filled with molecules, that great movements dominated its surface: whirlwinds, dust storms, hurricanes and tornadoes made up from the surface's loosened molecules.

31

The constant and speedy arrival of the Heat Entity through the Sun's beams on half of the revolving planet was greater than its escape. After striking and loosening itself from the inner tension of its beams, the Heat Entity increased its pressure, especially in that belt of the planet's surface which received its beams in the most direct-vertical manner. Its constant ever-increasing pressure—in the illuminated half of the planet—gave it during this half of the time, the superiority over the denser area of the Cold Entity, which continued to surround the planet's sphere. This superior power of its pressure against the resistance of the hardened Cold Entity, gave it the ability to expand and to thrust out the Cold Entity in the area of the planet's illuminated half.

As the Cold Entity is pushed aside under this pressure, it increases its hardness within the other half of the planet. This increased hardness gives it the necessary power and superiority to thrust the Heat Entity out of the non-illuminated surface.

Part of the Heat Entity which is thus thrust out escapes above, beyond the area and the atmosphere of the hardened Cold Entity, where it mingles with the mainstream of its own Heat Entity.

If all the Heat Entity which constantly beams down from the Sun unto the planet remained upon it for a considerable time, firstly, the Heat Entity would thrust out the Cold Entity from the planet's surface in an explosive manner; secondly, the planet's surface crust would melt into a liquid; and, thirdly, the Heat Entity molecules would start to burst and the entire planet would become aflame.

So that, this mutual thrusting out and expansion of the two Powers, that is, Entities, upon the planet's surface are the only *cause* of all winds, whirlwinds, storms, hurricanes, incoming and outgoing tides and ocean currents. All of the latter have their beginning, not beyond the atmosphere, but on the planet's surface: that is, in the atmosphere, in the waters, on the land, and in whatever place where one of the Entities has the greater power to expand and thrust out the other. Therefore, the velocity of winds and the change of their direction depends upon the change in this balance of power of one Entity over the other in any given area.

CREATION OF ATMOSPHERE AND WATER

The molecules which form the planet's surface are rough, unevenly round and grouped haphazardly together. Molecules must undergo a long period of refinement in order that they may be suitable as components of various kinds of matter, such as, air, water and for the creation and maintenance of living forms during the span of their existence.

The rotation and friction of molecules among each other in the stormy atmosphere, smooth out their roughness and give them more perfect rounded shapes.

The Sun's beams and all the fast rotating molecules prevent all the other molecules from linking together and consolidating in the atmosphere into a hard core substance. In the non-illuminated half of the planet's surface, or in places where a concentration of clouds of molecules screen out the Sun's beams because of their density, the Cold Entity—especially in the clouds—unites and compresses into groups those molecules which are inclined towards cohesion. These groups may be in the form of liquid, snow or ice, depending on the density of the Cold Entity and the pressure exerted by the Heat Entity in a given area. In whatever form they may be, the group of molecules which fall upon the planet's surface *becomes* "water" upon it. The fast rotating molecules which remain within the atmosphere *form* the "air" within it.

The refinement and sorting out of molecules continues in the water and in the atmosphere. The constant emanation of molecules from the loosening of the planet's surface increases the density and the size of clouds which in turn increase the frequency and length of falling moisture.

The water from this falling moisture, coming down from elevated parts and gathering into the deeper hollows of the planet's surface, loosens, crumbles and washes away in its path the loosened simple or compound molecules. Some water molecules, decomposed by the Sun's beams, evaporate into the atmosphere. Others wedge themselves into the planet's crust through their intermolecular spaces. Still others separate themselves from the water molecules and form the sediment under the flowing down or momentary still water.

The atmosphere's increasing density and the new presence of water on the planet's surface delays the decomposition of its hard surface by the Sun's beams.

The process of decomposition, refinement and sorting out of molecules and the creation of atmosphere and water from them lasts continually until the Sun's beams are stopped by the formation of a molecular crust on the Sun itself. Each molecule uses itself up through friction and loosening itself from its crust from the Cold Entity's surface. Those primordial molecules of the Heat Category which rotate the fastest are those who use themselves up the quickest and finally burst open. All refined molecules of the Heat Category—especially those who formerly had been part of vegetation—burst open even from the slight striking motion of the Heat Entity escaping from the nuclei of bursting molecules. On the other hand, it follows that all Cold Category molecules, because of their slower rotating velocity and ever thicker and harder crusts, consequently, prolong their existence.

The vanishing primordial molecules of the Heat Category are replaced by the original primary molecules emanating from within the exploding multiple molecules of the same category. However, the bursting, multiple molecules of the Heat Category are replaced by the refined, multiple molecules of the Cold Category, which through the partial tearing off of their crust, have lost the predominance of the Cold Entity particles in their crust over the Heat Entity particles in their nuclei; so that, upon losing this predominance, they now belong to the Heat Category. All vanishing molecules are replaced by other molecules from the constant decomposition of the planet's crust.

THE ORIGIN OF LIVING FORMS

With the beginning of the mingling and gathering together of molecules from both Categories, of various kinds and classes, within the water which fell and later gathered in hollows on the planet's surface, there probably began the creation of the first plant cells in the areas where the two Entities were found to be in the right proportion to each other and in a suitable state to create these plant cells, as well as in the areas where groups of refined molecules of both Categories, kinds and classes, were found to be predisposed towards the formation of certain plant forms.

One can surmise that in the creation of the first plant forms from primary molecules, there weren't enough loose and smooth (refined) molecules, especially of the Cold Category, nor a wide enough selection to form many varieties of plants. So that one can further surmise that these plants were of short duration and many died without reproducing themselves.

Each living form created on a planet, whether in the water, land or air, is a tool in the decomposition, sorting out and refinement of molecules, and an instrument in supplying itself and others with the necessary materials for their construction during the entire span of their existence; and after which existence, it itself becomes the material from which other forms are built.

No form—whether a plant or self-moving creature— originated its species by itself; nor was it created in a mature form all at once. The first forms of both genders initiating a species consisted in many try-outs in many places. The first

39

complete form of a species had its beginning within the embryo of the seed from which it issued.

No new permanent species originated from the cross-breeding of species.

The primordial molecules of both Categories linked together in simple, distinct groups—in the atmosphere, in the water or on land—are the first buds issuing seeds towards˙ the formation of the first forms of life on a planet. However, not all such primordial groups are suitable towards the formation of such seed buds. In order that a group be suitable to become a bud for certain seeds towards the formation of certain forms, it must contain within itself a certain quantity of primordial molecules of each Category and be so arranged that the primordial molecules of one Category are touching an equal number of primordial molecules of the other Category. This arrangement of these primordial molecules controls their reciprocal activity and maintains the required state towards the formation of a given form. Each of these groups, which is surrounded by various multiple molecules of both Categories, becomes the seeds from which are produced the forms of certain living species. On the other hand, those groups which are not suitable towards the formation of seed buds produce shapeless and fruitless algae in the water and on moist water-retreated lands. The first living forms, whether plants or self-moving creatures—whether in the air, water or on land—originated from shell enclosed seeds.

The outline of the form depended on how many kinds of primordial molecules of both Categories were involved within the seeds. Whereas the complete form of the species depended on how many specific and how many kinds of multiple molecules there were of each species within the seeds.

The time necessary for the incubation of a form within a seed, that is, from a perfectly finished seed up to the sprouting of its plant or the hatching of a self-moving form depended and depends upon the species, the size of the seed and the inner arrangement of the organism. If a given seed was formed from the smallest number of specific molecules and from the fewest kinds of each species—whether from primordial or from multiple molecules—then the complete form of the species also required the shortest time for its incubation than all the other living species, whether in the air, water or on land. But with the increased size of buds, the seeds also became bigger and their inner forms became more complicated. Hence, it takes accordingly more time for the Entities to sort out and arrange the molecules within the seeds into more perfect forms.

The Cold Entity within the seeds expands them with itself and keeps them in the same places. The Heat Entity, pushing through the Cold Entity in the seed, sorts out, moves and arranges suitable molecules into suitable places according to the respective outline and detailed form of the available supply and selection within the seed of the grouped primordial or multiple molecules.

So that, from the beginning to the end of each living being's existence, the two Entities are always and constantly active—without any interruption—in its construction, changes and final decomposition of its body.

With the advent and experience of each new life form, there occur more and ever newer kinds of more improved molecules. As more kinds of smoother and more improved molecules of the two Categories become available, so more perfect forms are being created by the two Entities in the plant world as well as in the world of self-moving forms.

Just as one Entity does not create from itself alone, so no living form can be created from molecules of one Category alone. Accordingly, a living form does not reproduce itself by itself alone—with the exception of those forms which contain in their embryo sufficient quantities of primordial molecules of both Categories and those multiple molecules of the kind from which these forms originated. The gender of any given living form depended and depends upon the predominance, within its embryo, of the primordial molecules of one Category over the other Category. Therefore, if the

41

Heat Entity in the nuclei of primordial molecules grouped in the seed's embryo predominates over the Cold Entity in its shell—in an otherwise equally balanced state of both Entities —then that embryo takes on the characteristic of the Heat Entity, which is the *male* sex. And consequently, if the Cold Entity predominates over the Heat Entity, then that embryo takes on the characteristic of the Cold Entity, which is the *female* sex.

In the form belonging to the Cold Category, at a given time, the two Entities select and deposit in pre-arranged places, special kinds of molecules necessary for seeds— from which it itself issued—for the reproduction and maintenance of its species and likeness.

The form belonging to the Cold Category accepts primordial molecules towards the formation of the embryo from a form of the same family belonging to the Heat Entity.

Therefore, whether it is in the seeds inside a living form —which are contained in a membrane with an inner opening and which are necessary for the supply of suitable molecules towards the development of another form—or whether it is on the outside in a tightly closed shell,—containing the necessary number and kind of molecules towards the creation of a given living being—the Entities are constantly at work selecting, dividing and arranging into suitable places the appropriate primordial molecules from which they create an outline of a form, which they further develop by means of the multiple seed molecules into a complete self-moving living being. In the arrangement of seeds towards the creation of a future form, the Entities maintain the most delicate balance towards each other: neither one, nor the other can overpower the other more than is absolutely necessary for the purpose. Otherwise, one Entity would remove the other from the seeds—even at its initial stage—and the entire construction of a form within the seeds would come to nothing.

Molecules are the first, smallest, individual forms created by the two Entities.

A molecule of each Category, according to its kind and

species, has a greater or lesser capacity for self-movement or adherence. If by any chance, molecules of a different kind would enter into the seeds—whether primordial or multiple, but not belonging to the form of a given species—or if there would be too many of a certain kind, or too few of those necessary for a given form, then it is probable that from such a seed no form would issue, and if it did, then probably it would be deformed. As a certain kind and species of molecules are harmful to self-moving forms, it would follow that if they entered in some way into the latter's seed or into an already developing form and united with the molecules of that given form, then they would act unfavorably: either prevent its development or cause its complete disappearance.

So that, plants of the Heat Category as well as plants of the Cold Category can propagate themselves perhaps because plants of the Cold Category cannot have such direct interaction with the Heat Category molecules of its species, as have self-moving forms, which therefore are the *start* of the multiplication of species. Because of this, at an appropriate time, the two Entities prepare places for the seeds within plants of both Categories. And these places also serve as a reservoir for primordial molecules for the use of the second plant Category of the same species, as well as an intercepting area of primordial molecules belonging to the second Category towards the use of the first Category's embryo. Therefore, at the time of the plants' mutual fecundation: the Cold Category plants intercept the primordial molecules of the Heat Category of the same plant species; whereas, on the other hand, the Heat Category plants intercept the primordial molecules of the Cold Category (of the same plant species) towards the formation of an embryo of the same species. For, as we have already explained, the Cold Category primordial molecules are cohesive by nature; whereas, the Heat Category primordial molecules' main characteristic is their high rotating velocity.

From time to time, the Heat Entity increases its inner tension (with the help of the Cold Entity and easy splitting

43

molecules) within the growing forms of both Categories in order to direct and compel them to unite for the purpose of transferring the primordial molecules for the formation of an embryo from the form belonging to the Heat Category to the ready, awaiting seed in the form belonging to the Cold Category. This occurs mostly when the reservoirs of both Entities are full—the Heat Entity reservoirs containing the primordial molecules of both Categories designated towards the formation of an embryo; and the Cold Entity reservoirs containing multiple molecules for the formation of seeds. Before and during the process of transferring the primordial molecules (destined towards an embryo) from the Heat Category form into the Cold Category form, the Heat Entity increases the pressure within the Heat Category being to the point necessary for this transference.

The Cold Entity increases its stiffness in order to take over the direction of the seeds within the Cold Category being so as to retain in groups the passing primordial molecules while preventing their mingling with the other molecules from the body of that given form.

THE SOUL

From the moment that the Heat Entity, with its primordial molecules destined towards an embryo, enters into the seeds of a body of the Cold Category, there begins within them the construction of the complete organism and the beginning of the fusion of the two Entities present within them into a new, *homogeneous, individual* state, which from now on we shall call SOUL. Since the Creators—in the literal sense of the word —are these two Entities who have mutually opposed characteristics, therefore, the fusion of this duality into *one, individual* and permanent state—or "soul"—requires the following conditions: firstly, the organism must be self-moving as only within it can the uniformity of this fusion (or soul) be achieved; secondly, the organism must provide suitable intermolecular spaces where the Entities can fuse effectively; thirdly, it must contain rotating molecules in order to mingle and fuse the two Entities together and provide movement to the organism. If these conditions are met, then the soul completely permeates all intermolecular spaces which completely surround all molecules in self-moving beings. Otherwise, if the self-moving being—especially land mammals who are made up of the greatest number and kinds of the more refined and smoothed out molecules and who have the most complex organs—were not permeated with this permanent, individual soul, but only with the original Entities, then they would not have within them a well regulated, permanent temperature. Their temperature would then change within them according to the outside temperature of the surrounding Entities and would fluctuate greatly according to which En-

45

tity in a given time and place would have the predominance over the other.

Therefore, the "soul"—or the fusion of the dual presence of these two Entities—prevents, to a certain degree, the invasion of a given body—whether in its interior or exterior—by the dominant Entity.

So that, during the entire span of this body's construction and existence, all instincts—found in that given body and its organs which guide them towards the acceptance, selection, reconstruction, transference and placement of molecules in their respective places—are always directed by both Entities, especially to a greater degree, by the Heat Entity.

The soul, within a self-moving being which has just issued from its seed into the visible world, is completely ignorant. Such being the case, all movements which are necessary to a self-moving being, especially during the infancy period, are executed directly and knowingly by the Entities without the soul's knowledge of its participation. The pressure of one Entity or the other, or of both, inside the body, propels the soul in the direction of those parts of the body which have to be moved.

The soul, thus properly propelled towards the performance of the indicated movements required in certain parts of the body, moves them still unconsciously in the initial stage of its existence, but eventually with a certain degree of knowledge in the period of its further development and its acquaintance through the senses of the visible world surrounding it as well as with the needs of its own body which are necessary towards the maintenance of the appropriate movements of its being in the course of its existence.

Therefore, if there should be any physical pressures exerted upon the body by material objects or any other living being—whether within itself or outside—the soul counteracts them by moving the body into appropriate, though involuntary, instant reactions.

In spite of the fact that the soul has the characteristics of both Entities—sight, hearing and feeling throughout its being—nevertheless, when in a body it can only see, hear and feel through the respective senses located in the body and which are connected by means of the permeated intermolecular spaces surrounding the sensory centers. That part of the soul which fills up the brain's intermolecular spaces makes notations on the brain's molecules—as on a film—of all pertinent observations on visible objects, voices, external and internal touching which can be grasped by the senses. Through frequent review of these notations on the brain's molecules, the soul informs itself on: how to conduct itself; how to provide for the necessary needs of its body; and how to protect it and maintain it unharmed, and alive as long as possible. And that entire process is given by the highest Powers, that is, Entities—though it is universally called intuition or instinct.

At the moment that the completed form of a self-moving being issues from its seed into the visible world, the soul within him has become an *indivisible whole*. Also, as the body increases in size so the soul within it continues to develop. But when the body has attained its final form, then the soul within it also ceases to expand.

The soul can never become stiffened into a hard substance as the Cold Entity can be within the molecular crusts; nor can it be contracted into a visible form such as the Heat Entity because it contains within itself the fused opposite characteristics of both Entities: that is, the hardening of the Cold Entity and the loosening of the Heat Entity. Also, the soul cannot become decomposed, or divided into parts, because it is not created from molecules, but *directly* from both Entities within the self-moving being into a homogeneous, individual whole and permanent state of being.

The soul's size depends on the size of the body in which it was created and how large is its form when the soul leaves it.

The normal state of the soul in a normal, self-moving (body) form, when there is no pressure within or without by either of the two Entities, is equivalent to the balance

47

between the fluctuating tension of the Heat Entity and the hardening state of the Cold Entity. This state of the soul is always advantageous and positive; it never deteriorates into the negative side of dilution or diffusion, whether during the time of its presence in any self-moving form, or after it has left it.

These two life-giving Powers, that is, Entities, *resemble* each other the most in their *nature,* but also *differ* the most— since they are opposite to each other—in their *characteristics,* more so than any of their creatures.

There never was, is or could ever be two creatures absolutely identical in all aspects, in whatever was, is or will be created, even though they would be similar, such as two primordial molecules belonging to the same Category and class.

Therefore, even souls differ among themselves: one soul can contain more of both Entities than another soul; while another may have more or less of either Entity than another soul, and so forth. . . .

Each soul belongs to that Entity of which there is more of within the soul, irrespective of what Category the form (body) belonged to within which this soul was created. Since there are *two* Creators, therefore, all the souls which they create belong to either of the two Categories.

Each self-moving form, from the smallest to the largest, which is created on a planet—whether in the air, in water, on land or within the earth—which is provided with a brain and organs of sight and hearing has within itself a permanent and complete soul.

But those forms which are unable to move themselves from place to place, or who do not need to be moved, and who have only one sensory feeling, then these forms do not have within them this individual soul. So that, whatever limited movements are necessary for them in order to feed themselves or for their protection are executed within them directly by the two Entities.

"Souls" are the real children of the real Creators of all

creation. Whereas bodies of self-moving forms are only temporary shapes and exist merely so as to have "souls" created within them. And the bodies' remains serve only as material towards the formation of the future forms of living beings.

But, the "soul" upon leaving its body becomes cognizant of itself and knows that it is still the same "I" which formerly was the "I" inhabiting the body. Also, the soul retains its original size and knowledge, plus the newly gained knowledge of from whence it came, that is; that it was *generated directly* from the *Creators* themselves . . . and that it will endure for a long, long time, and perhaps even forever!

SUPPLEMENT to MSS from NOTES

Therefore everything that exists in the universe—except the soul—and which had a beginning, will cease to exist. However Beings, who did not have a beginning, but whom we know exist, then these will exist forever.

But everything which had its beginning and has a visible form—although perhaps not seen by the naked eye—is created from something and by someone. But only "Beings" who did not have a beginning and who are without a form can create all forms on the planets and throughout the universe.

The following is universally believed: that Heat is created from "Fire"; that fuel, whether in solid, liquid or gaseous form, creates fire; that easy combustive materials provide fuel; that the earth's crust provides materials; and that the earth's crust is created from atoms! . . .

But when Heat escapes from the state of fire, and the fire in turn escapes from what concealed it, then nothing remains except the debris from the split atomic crusts; and that will exist only as long as the Cold shall not dissolve or loosen from it. For the more Heat is compressed, the faster it dissolves. Whereas, the more hardened Cold is, the longer it takes to loosen.

MIXED FIRE

Mixed fire, that is, the ordinary fire which we see on our Earth coming out of the molecules from the combustion of any fuel, mixes with hard-splitting molecules and with the debris of broken down crusts of molecules.

PURE FIRE

The fire emerging from split molecules can be pure when the nuclei escape through splitting molecular crusts and enter into the gaps of uniformly laden molecules of one kind and of approximately the same size before they let go of their inner compression.

The constant flow of nuclei into the gaps of a given material creates within it what is called "an electric current." So that, two or more currents connected to the two opposite ends of given materials, create a pure and visible fire. Its brightness and force of impact depend on the increase or decrease of the flow of nuclei into the gaps of the given materials.

53

PURE FIRE IN LIGHTNING

The Heat Entity, which concentrates upon the Earth's surface from the loosened Sun's beams, is thrust upwards by the Cold Entity which loosens itself from the molecular crusts from the Earth's crust.

When the Heat Entity is thrust upwards, it brings about the hardening of the Cold Entity and at the same time it increases the pressure within itself which in turn compels it to decrease its volume.

If it should happen that certain sections of the Heat Entity would become surrounded and detained by the Cold Entity, then those Heat Entity sections compress themselves into visible "lightning" and pierce through the hardened layer of the Cold Entity which immediately dissolves itself.

The Cold Entity hardens the most in the area where it touches the Heat Entity.

In conclusion of the above, one can say that no fuel— whether from solid or fluid substances, or from any volatile gases or atoms—creates fire on planets, suns or lightning. For fire is always ready within the atoms' nuclei, the planets' cores and visible in suns.

So that we see from this that only Heat can create visible fire; and not fire create heat. Because everything vanishes, except Heat and Cold which remain forever.

All forms on each planet and in the entire universe— beginning with atoms and ending with suns—are merely temporary conservation measures to preserve that from which they were created.

So that, according to the above, one cannot agree with the supposition that certain gases are burning within the suns. Also one could not agree with the further supposition —as some maintain—that "lightning" is the burning of gases.

For in lightning as in the Sun, the pure fire which we see, is created directly when the Heat Entity is surrounded by the Cold Entity and the Heat Entity consequently compresses itself into a visible state!